10 things every kid should know about GOD

I0163462

10 things every Kid should know abōut GOD

by Tina V. Bryson

Illustrations by Mykle Lee

Leeway Literary Works
Published by Leeway Artisans, Inc.
9468 Pep Rally Lane, Waldorf, MD 20603

Book & Cover Design Mykle Lee

ISBN: 978-0-9823349-4-2

Copyright © 03/2010 by Tina V. Bryson.

Jesus Letter Copyright © 2009 by Jason Byerly, Southland Christian Church,
adaptation used by permission.

No parts of this book may be reproduced or utilized in any form or by any means, electronic or mechanical,
including photocopying, recording, or by any information storage or retrieval system, without permission from the
Publisher. All inquires should be addressed to:
9468 Pep Rally Lane, Waldorf, MD 20603.
ALL RIGHTS RESERVED

First Edition
Printed in the United States of America.

King James Version Public Domain in America
New King James Version® Copyright © 1982, Thomas Nelson, Inc. All rights reserved.
Scripture taken from the HOLY BIBLE, NEW INTERNATIONAL VERSION®. Copyright © 1973,
1978, 1984 Biblica. Used by permission of Zondervan. All rights reserved.
Scripture quotations are taken from the Holy Bible, New Living Translation, copyright 1996, 2004.
Used by permission of Tyndale House Publishers, Inc., Wheaton, Illinois 60189. All rights reserved.

Contents

Dedication

I dedicate this book to the children and families of Consolidated Baptist Church. Thank you for sharing your lives with me, and allowing me the true privilege to invest my life with you. May God continue to help you Know Him, Grow in your faith, Show God's love, and Go wherever He sends you to tell what He has done in your life.

INTRODUCTION

When I was a kid, I used to ride my bike to our church, sit on the steps, and talk to God. Looking back, I think I felt that it was the place where I could be closest to God.

As I was growing up and going to church, I didn't realize that He didn't want to be a part of my life just at a building. I also didn't really know that Christ had died for me, that He loved me, and that **salvation** meant more than just being a good person. God wanted to be a part of my whole life. He wanted to live in my heart. Salvation was about having my own, very personal, very special relationship with God through His Son, Jesus.

Now, as a mother, I don't want my children to love God just because I love God. I don't want them to choose Christ just because I did. I want them to know that God loves them and that He chooses them. God wants to have a personal, special relationship with them through His Son, Jesus. I want them to know that God has an awesome plan for their lives. There is a divine purpose for them being here on Earth. I want you to know these same things.

When I started teaching in children's ministry, I was reminded of when I was young. I would go to church and learn Bible stories, but I didn't always see how those stories related to me. My favorite Bible story growing up was the Prodigal Son. It is still my favorite story today. Then, I liked the fact that the son was able to come home, and his father was standing on the road with his arms open wide to receive him. Now, I love the story, because I understand what it truly illustrates. The father loved his son so much that it didn't matter what he had done. He was always waiting for him to return. When the son chose to come back to his father, they could start a new relationship.

Like the son in the story, you can choose to come to your Heavenly Father, God. That is what it means to be a Christian. God is waiting for that chance to have a personal, special relationship with you.

The purpose of this book is to teach you some important things about God to help you grow in your relationship with your Heavenly Father. I hope that this book will show you how much your Heavenly Father truly loves you and wants to be a part of your whole life. Like the father in the story, God has His arms stretched wide to receive you.

He is waiting...not just for anyone, **God is waiting for YOU!**

GOD SAVES IN AN INSTANT... THE MOMENT WE PRAY

When I was a child, about the age of 10, a classmate asked me if I was saved. I immediately said "Yes," because I could tell from the question that being saved was something that I should want to be. However, I had the feeling that being saved was something that she thought I was not.

Now I must confess that at the time, and for some time later, *saved* was just a word I had heard at church. I didn't really know what it meant. *Saved from what? The devil? Going to hell? Being a bad person? Having bad things happen to me?* Not only didn't I know what it was, but I didn't really know how to *be* it.

And if you're like me...you may have the same questions.

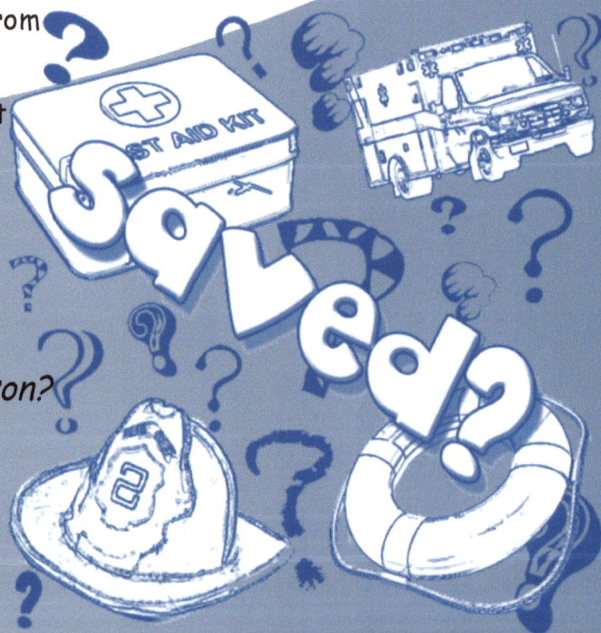

What does being saved mean?

Let me explain it this way. When I was 14, I went to a pool party at my friend's house. Our softball team was celebrating winning the playoffs. We were headed for the state championship. I couldn't swim, but I still liked to play in the pool.

I would always start in the same, sitting on the side with just my feet and legs dangling in the water, or I might start by sitting on the steps with the water meeting my waist. Without fail, I would slowly become comfortable in the water, and would start venturing in further. I knew enough to know that I had better stay in the shallow end.

On this particular day there were lots of pool toys being tossed around and floating on the water. I had a particular interest in one and I played with it for quite some time. At one point, one of the boys at the party took it and I began to chase after him. I wasn't worried because I was still in the shallow end, and I could feel my feet on the bottom of the pool. As I chased a little further. I knew I was getting closer to the deep end, but I wasn't worried because I could still feel the bottom of the pool. I went further and further, still chasing the toy. I was so focused that I didn't notice how far out I had gone until...SPLOOSH!

The bottom of the pool seemed to disappear. I was in the deep end and totally under water. I could still see the boy pulling the object I had been chasing, but he just kept swimming, never noticing that I was drowning.

A Thief Saved in an Instant

"But the other criminal rebuked him. 'Don't you fear God...we are getting what our deeds deserve. But this man has done nothing wrong.' Then he said, 'Jesus, remember me when you come into your kingdom.' Jesus answered him, 'I tell you the truth, today you will be with me in paradise.'"

- Luke 23:40-43, NIV

I began to panic as I sank. I whipped my arms, and tried to grab anyone that I saw swimming by. I felt water come into my nose and mouth as I continued to kick my arms and legs wildly. At that moment, I knew that I was surely going to die. I did not doubt it because no one saw me in distress. I was in a pool surrounded by people who were all enjoying the party and having fun, yet I was dying, right in the middle of it all.

And then something inside me told me that I had to calm down. I had to stop kicking because I was only making death come faster. I stopped panicking, and reached out my hand. Suddenly, a friend reached down from the side of the pool, gripped my hand in hers, and pulled me up and out of the pool. In that moment, that split second that she reached for my hand, she rescued me. In an instant, I went from death to life. I was *saved*.

And that is what it means to be saved. In an instant, God takes us from death to life. Just as my friend reached down, God is reaching down to you. He is your friend, and the moment you take His hand, you are saved.

When I was that child, I did not know how God saved people, but I discovered that it is very simple. As I kicked and whipped my arms that day in the pool all I did was cause myself to sink faster. Yet, being unable to swim, I would not be able to get out on my own. Reaching my hand up was my way of calling out, "Help me!" We can do the very same with God.

Many times we do say, and even think things that hurt God because our hearts are disobedient to Him. Disobedience to God is a lot like drowning because we're sinking in a pool of sin that only leads to death. The only way to be rescued from this pool is to seek God's salvation through prayer. When you pray for salvation you are calling to God, reaching out to Him with your words and your heart and saying, "Help me." And when you call out to Him, He will answer. In that moment that you pray that prayer, **He saves you!**

GOD CHOOSES YOU

When I was growing up, I liked to play kickball. One of the best feelings in the world is getting picked as a teammate. On the other hand, one of the worst feelings in the world is *waiting* to get picked. I have been in both places. I have been chosen as the captain by the teacher, and I have waited for friends to pick me to be on their team.

Occasionally, the captains did not want me on their team, for whatever reason, and I stood while name after name was called. Each time one of the captains prepared to call someone, I hoped to hear my name. And each time someone else was called, I felt rejected. As the group got smaller and smaller, I felt smaller and smaller. The very worst feeling was not getting picked at all...being the last person left so a team had to take me whether they wanted me or not.

The good thing about God is that He's the captain who wants everyone on His team. The even better thing is that God's team is never full. In fact, the best thing is that He loves us so much that He gives us the choice to be on His team.

God pursues each of us because He loves us so much and wants us all to have the opportunity to say yes and join His team. And *He never allows us to be the last pick.*

God wants YOU on His Team!

Let me explain it this way. Some years ago, we met some friends with three children of their own. They were a family of five—mom, dad, one son, and two daughters. Each child was very special to them because God had given each child particularly to them. Their children knew, and everyone who met this family knew, God had chosen to bless them in a very special way. They loved each other very much.

This mom and dad loved God so much. And God had given them so much. So they wanted to help other children to have a family too. They went to another country and met several children who didn't have parents around to take care of them. This time they chose to bless three other children in a very special way. They picked one boy and two little girls to be a part of their family.

They didn't just bring them to their house, give them a bed and say, "you better be glad that we brought you here." No, not at all! They adopted those three children, who then became part of their family. The three new children had the same rights and privileges as the three who had been there all along. The mom and dad did not make any difference between them. This family of five was not too full to become a family of eight.

Pursued and Adopted

"God demonstrates His own love for us in this: While we were still sinners, Christ died for us."

- Romans 5:7-8, NIV

"For all who are led by the Spirit of God are children of God..you received God's Spirit when He adopted you as His own children. Now we call Him, 'Abba, Father.' For His Spirit joins with our spirit to affirm that we are God's children. And since we are His children, we are His heirs."

- Romans 8:14-17, NLT

God is the same way. He chooses us! He picks each one of us to be a part of His family. True, He already has one Son named Jesus, but He adopts us and makes us His own. God does not make a difference between us and Christ. Romans 8:13 reminds us that we are heirs of God and co-heirs with Christ.

That same wide grin that came across my face when I was picked to be part of the kickball team, comes across my face now…but even wider. I am so happy because I know that God, who created everything that I can see or imagine, He chose me, and **He chooses you, too! WOW!**

GOD GIVES US NEW LIFE

When I was a kid, we had this gigantic open field not far from my house. In the summertime, we would ride our bikes to the field and start a game of kickball. Sometimes, as we were standing out there, the summer sun beat down on us, so much so that even our baseball caps could not keep our heads shaded. Sweat ran down our backs, down the sides of our faces, even down our arms and legs. We would swat the bugs away, rub our dirty hands across our foreheads, and hope a cool breeze would blow. Someone would come up to home plate, worn out from the summer sun, and wait for the pitch from the mound. As the ball would come rolling, something would go wrong…a bug buzzing near, sweat dripping into the eye, a welcome breeze distracting us…and instead of kicking the ball the right way, the ball would roll past and the kicker would yell out, "DO OVER!"

Sometimes, that's how we feel about life. *We want a "do-over,"* *a second chance* to do something and try to get it right. However, there are very few things in life that you get to "do-over." You can't do-over a test if you get a grade you don't want. You can't do-over a spill caused at the kitchen table. Even accidental things, like hurting someone's feelings, sometimes can't be done over. But *God gives us a chance* to try *to get it right* the second time.

God gives us a "Do-Over?!"

I was teaching a class one day, and a student raised his hand and asked a question. *"Is new life in Jesus like when you're playing a video game and your character gets killed, but you get a new one?"* His parents tried to make him be quiet, reminding him that we were in church, and we were talking about Jesus.

But after a few minutes, it occurred to me that getting new life in Jesus is kind of like getting a new character in a video game. In most video games, you get a character and you have a goal. As you play the game trying to reach that goal, you have to make choices. Some choices are good, and some are bad. Bad choices might be deciding to go one way instead of another, or not making the right move and getting clobbered by your opponent. In the end, the consequences of all the mistakes you have made in the game catch up with you, and that is the end of your character. Sometimes your character gets killed or something else, equally bad, happens. But instead of the game being over, you get another character, a second chance, a "do-over." Well, that is exactly how new life is in Christ.

Christ's goal for every person is heaven. He knows we can't get to heaven without Him. As we go through life making choices, doing this thing, or that thing, the choices we make on our own do not get us into heaven. Eventually, the consequences of all the choices we make add up. We may not get killed, but something equally bad happens. But then *God gives us a "do-over."* We ask God to forgive us, and we get a second chance. God takes away that old person that we were, and we get a new character. We get a chance to

make the right choices.

2 Corinthians 5:17 says, "Anyone who believes in Christ is a new creation. The old is gone! The new has come!"

Just like a video game...but with one big difference. When you're playing a game, and your last character messes up, you get a "Game Over" message. When we get this new character, this new life is ours forever. We have a chance to live for God forever, and there is no "Game Over" because even when this life ends here on Earth, we get to go to heaven and live with God forever. Now that's a great **"Do-Over!"**

GOD HONORS TRUE REPENTANCE

One day I was very surprised to hear my first grade son using words like amble, gullible, vast, and crave. I had not realized that he was learning such extensive vocabulary words at school. I was more impressed by the fact that not only could he pronounce the words correctly, but he could also use the words in proper context. He understood what they meant, and how to apply them to things in his own life.

All of this was very exciting to me because I know that words provide a way for us to communicate very complex ideas or very simple ones.

Well, *God uses words*, in the Bible, *to communicate ideas* to us. Sometimes they are very complex ideas, like redemption. And sometimes they are very simple, like love. One word that can sometimes be misunderstood, because it seems simple, is repentance.

When we hear the word repent, it is usually being used about someone who has done something wrong. We expect some kind of response from them. Namely, we are looking for them to apologize, to show that they are sorry for what they have done wrong.

I'm Sorry... x 10

"And Pharaoh sent for Moses and Aaron, and said to them, 'I have sinned this time. The LORD *is* righteous, and my people and I *are* wicked. Entreat the LORD, that there may be no *more* mighty thundering and hail. I will let you go... So Moses went out of the city from Pharaoh and spread out his hands to the LORD; then the thunder and the hail ceased. And when Pharaoh saw that the rain, the hail, and the thunder had ceased, he sinned yet more; and he hardened his heart ....neither would he let the children of Israel go."

- Exodus 9:27-28, 33-35, KJV

If you've ever had someone do something wrong to you...like your sister hit you, your brother broke something of yours, or a friend did something but you got in trouble for it...you wanted that person to say to you, "I'm sorry." And sometimes, that is all you're looking for, an apology. But if that same person who apologized did the same thing or maybe even worse to you another time, over time, you would wonder if the apology was real.

Well, repentance is more than an apology to God for the things we have done that He told us not to do. Repentance is a word that God uses to teach us about our relationship with Him. If you look it up in the dictionary, it really means "to turn from sin."

So what is true repentance?

Picture it this way. Imagine yourself walking in one direction, and you're just doing your own thing, going your own way. Then you realize you're going the wrong way and make a decision to turn around, right where you are, and walk in the opposite direction. You are turning from one path and choosing an opposite path. That is what repentance means.

Repentance happens when you turn from that one thing and do the opposite. You don't just feel sad about it or say you're sorry. Realizing you've done wrong, you turn around and go the opposite direction. And, perhaps, you make things right with the person you hurt or work to make sure someone else does not make the same mistake.

When you really love someone, you don't want to keep doing something that you know will hurt the person you love.

When we make choices that don't please God, because we are doing our own thing, sometimes it doesn't even bother us. So we say we're sorry, and then we ask God to forgive us, but then we just go on doing our own thing. That's not true repentance.

True repentance to God is realizing that you have done something to upset God, and understanding that because you love Him, you don't want to do that ever again. True repentance is only possible through the power of the Holy Spirit. We are not able to do it on our own. It takes God's power in us to live a life that pleases Him

He is faithful and just to forgive...

The best thing about repentance is that when we really start loving God and wanting to please Him, He helps us recognize the things that we may do or say that make Him very sad. Repentance is a word that God uses to help us understand our relationship with Him.

So as you read your Bible, talk to God, and learn how God wants you to live, sometimes you may do something and you realize that you have sinned against God, you have hurt God. Then, you have a choice to make. You can choose to just say you're sorry, and keep on doing those things. Or you can repent. You can make a choice to go in the opposite direction. If you used to tell lies, now you tell the truth. If you used to steal, now you give to others. If you used to say bad words, now you say things that encourage other people and make them feel happy.

Romans 6:1-2 says:

"Well then, should we keep on sinning so that God can show us more and more of His wonderful grace? Of course not!"

We know that God will forgive us when we ask Him and we are truly sorry for what we have done. But we shouldn't just keep saying we're sorry and keep doing the same thing over and over. When we are really sorry, and we ask God to forgive us, our goal should be to repent, to turn away and do the opposite. We know that God loves us very much, and we don't want to take advantage of that. We wouldn't want someone to treat us that way. So, the next time you remember that you have done something, said something, or even thought something that made God very sad, don't just say you're sorry....***repent***.

SALVATION IS GOD'S FREE GIFT

One day, my son forgot his lunch box, and his friend bought his lunch. He did so just because they are friends, not expecting anything in return. He even told my son directly that he didn't want the money back. Still, when I picked my son up from school, he was trying to decide how to repay his friend.

I reminded him that *sometimes people just want to do something for you*, and they don't expect anything in return. When you go out of your way to do something to pay them back, you actually take away a little something from what they did for you.

Think of it like this. Generally when you go to a birthday party, you bring a gift or a card and you put it on the gift table. You are so excited to see your friend's face when they open the present that you brought. You watch them pick your gift from the table, and as they tear off the paper, you get more and more excited because you know they are going to love your gift. But what if that birthday boy, or girl, came out before opening the gifts and gave everyone money to pay them for the gifts that were brought? Now it is not really a gift at all, because you were paid for it. They really bought their own present. You just got it from the store.

How would you feel when after they paid you the money, they opened the box and talked about what a great job they had done in getting this gift? Well, God knows exactly how that feels.

God expects nothing in Return.

You see, God offers us salvation as a free gift. He sent Jesus, His Son, to die in our place, to take the punishment for all of the things that we would think, say, and do in disobedience to what God told us to do.

God offers us a free gift, but sometimes people want to return the favor. They want to pay God back. So they go to church more often than most. They put lots of money in the offering plate. Sometimes they join a ministry like the choir or ushers. They do all of these things, not just because they love God, but because they want to show Him how good they can be. They plan what nice things they can do for God, to return the favor.

But if we have to do all those things to earn God's love or to earn salvation, it is not a free gift at all. We have worked for it.

God puts it this way in the Bible. Ephesians 2:8-9 says, "God saved you by His grace when you believed. And you can't take credit for this; it is a gift from God. Salvation is not reward for the good things we have done,

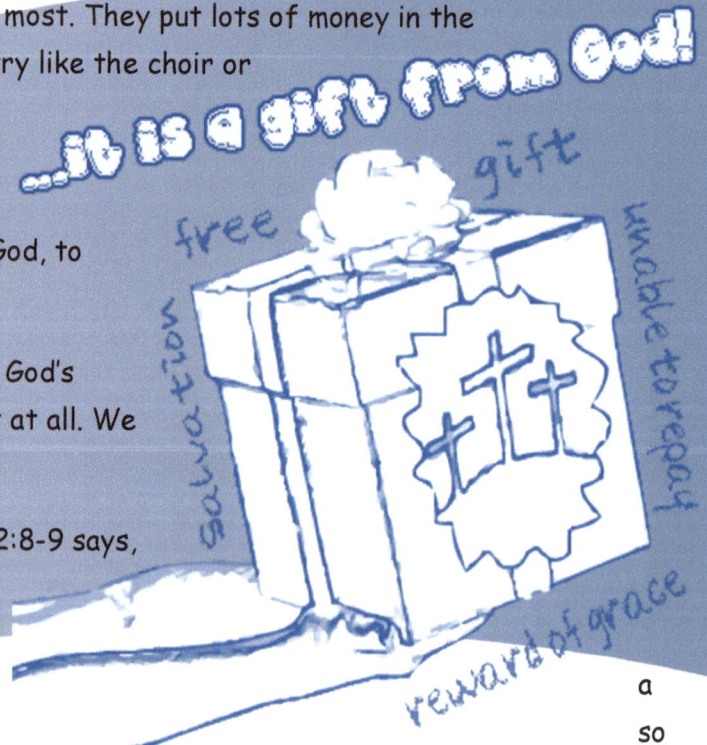

...it is a gift from God!

free gift

salvation

unable to repay

reward of grace

a

so

none of us can boast about it."

God loves us so much that He's already done all of the work for us to get to heaven. We need to have a grateful heart, and **accept what God has done for us**.

ONCE YOU ARE IN GOD'S FAMILY, YOU ARE ALWAYS HIS

We have four children in our family. The oldest is already a teenager, but I have not forgotten what he was like as a baby or even as a toddler learning to walk, talk, and do things for himself.

Some days we have so much fun together, and other days not so much. You probably have days like that…those times when mom or dad is giving you a lecture about cleaning your room, doing your chores, or staying on top of your grades. You know the kind where parents are talking, but all you hear is "blah, blah, blah, blah," and as the parent gets more determined to make the point, your eyes start to roll or your mind starts to drift to "what's for dinner," "I'm going to miss my favorite show," "will this ever end?" Those are the easy times.

There are more difficult times in the lives of parents and children that pull at the bonds of what makes a family, what holds a family together. Those are the times when parents are so disappointed, but they hope they can give you some guidance, direction, or words of wisdom that will help you make better choices that don't have

negative consequences. But no matter how difficult those times may become, one thing can't be changed...*you are always their child*. And it's the same way with God.

God is your Heavenly Father, Forever!

*L*et me tell you a story about when I was in high school and I had my own car. One night, I stayed out late while hanging with a friend and I came in about 7 a.m. I didn't think about my mom, how worried she'd be wondering where I was. I couldn't even understand, at that time, how much I had disappointed her. She was probably at the breaking point. But at that moment, even if she had decided that she'd had enough, she was going to disown me, and be done with it, one thing my mother could never change was that I was her daughter. She would always be my mother, no matter what. It didn't matter what she said, what she did, how she felt... She could not change the fact that she was my mother.

She took my car away, and I could only go to school and church. I could have rebelled and stormed out of the house. I could have jumped in my car and sped away, never to be seen or heard from again. And you know what? I would have still been her daughter. Even if I went to another place where no one knew me, or even if I changed my name, I was still her daughter. I couldn't change the fact that she had given birth to me.

You're in GREAT Hands

"My sheep listen to my voice; I know them, and they follow me. I give them eternal life, and they shall never perish; no one can snatch them out of my hand. My Father, who has given them to me, is greater than all; no one can snatch them out of my Father's hand. I and the Father are one."

- John 10:27-30, NIV

God looks at us in the same way. Indeed, Paul tells us in Romans that he is:

"absolutely sure that not even death or life can separate us from God's love. Not even angels or demons, the present or the future, or any powers can do that. Not even the highest places or the lowest, or anything else in all creation can do that. Nothing at all can ever separate us from God's love because of what Christ Jesus our Lord has done." ~ Romans 8:38-39

That is so awesome to know that even when I mess up, when I do the wrong thing, even when God is angry with me, He will never disown me. I am always His. God says in Hebrews 13:5, "I will never leave you. I will never abandon you." WOW! I love that! Once you are a part of God's family, **He will always be your Daddy, no matter what.**

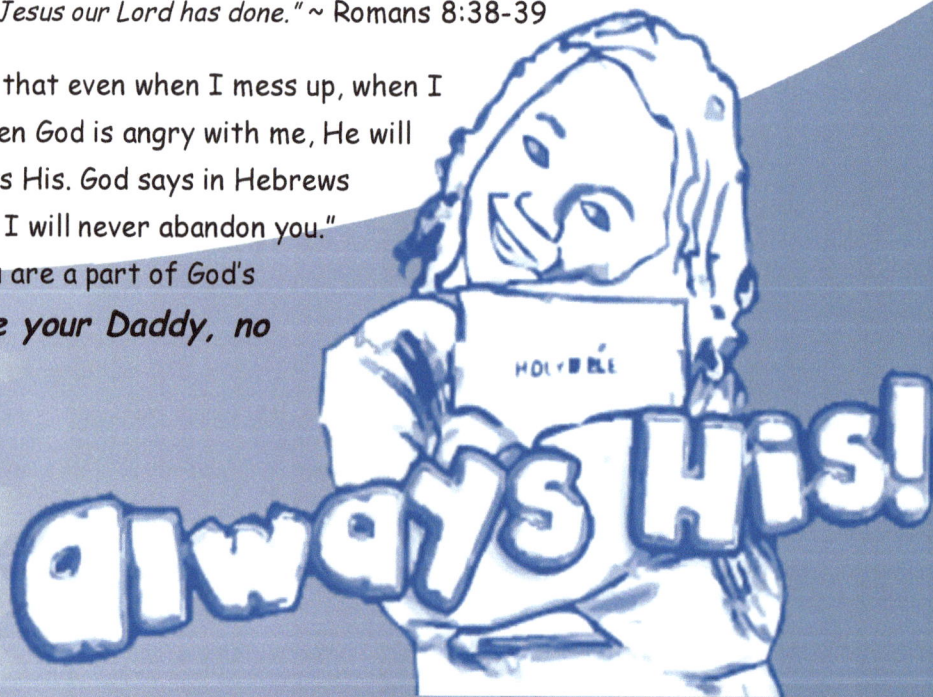

GOD DESIRES YOUR INWARD COMMITMENT AND YOUR OUTWARD EXPRESSION

There are moments in life when people tell you *"it's about that time for you."* We all know when it is time to do something…like go to bed, eat, or take a bath. But sometimes when your mom or dad, grandparents, or even your teacher says "it's about that time for you" it could mean that you have reached a certain age to do something different or new. Sometimes parents will warn that you are too old to still be sucking your finger. It's time to give that up. Or that it's time you start earning allowances. Maybe it's a grandparent who reminds you that you've reached the age where you need to help your parents out more.

There are many things that we learn to let go of as we grow up, and many more things that we learn to embrace as we go from being toddlers to school aged, from tweens to teens, and to young adults. And in all of that growing up, sometimes kids are told that it's time to

join the church, or it's time to get baptized.

Here's a story I'd like you to consider. I hope it helps you decide if that time is *the right time for you*.

Two months after graduating from college, I met the nicest guy in my Sunday School class. We had a lot of things in common and I thought he was pretty special. He became my boyfriend. Then just eight months after we met he asked me to marry him. I had prayed long and hard about marriage even before I met my boyfriend. It was important for me to look into my heart and know if it was the right time. And it was. I said yes and it was truly one of the happiest days of my life. After he put that diamond ring on my finger, I just couldn't stop looking at it, marveling at it. I was thrilled to show everyone because it proved that we were ready to express our commitment to one another. I was even more thrilled on our wedding day, when we said our vows in front of our friends and family, showing the whole world that he was my husband and I was his wife.

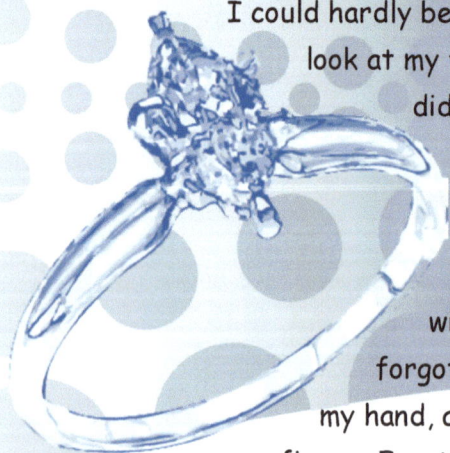

I could hardly believe that I had gotten married. From that moment, anyone could look at my finger and know that I belonged to someone special. Sometimes, I didn't even wait for them to look. I would just hold up my hand, and show my beautiful ring. I was a married woman.

Not long after we were married, we were on our way to church, and I took off my ring to put lotion on my hand. Still chatting with my husband, I did not notice as I got out of the car that I had forgotten something. Later, during the church service, I looked down at my hand, and realized I had lost my ring. I had forgotten to put it back on my finger. Frantically, we searched the church, the car, and the parking lot. We never found the ring.

It took some time to get another ring, but even without it I was still married. The ring was not the marriage. It was just a symbol, or outward sign of an inward commitment. The marriage was real because my husband and I made a commitment to each other. We said vows before our family and friends, and most importantly we said vows before God, that we loved each other and wanted to be together for a lifetime. The marriage was sealed in our hearts. So even though I no longer had the sign, the decision that I had made on that day was still true.

It's an outward symbol of an inward commitment.

*J*oining the church, or even being baptized, is just like my wedding ring. It is an outward symbol of an inward commitment. It is to show people a sign of a decision that you have made in your heart. People

Express Your Love

" And we can be sure that we know Him if we obey His commandments. If someone claims, 'I know God,' but doesn't obey God's commandments, that person is a liar and is not living in the truth. But those who obey God's word truly show how completely they love Him. That is how we know we are living in Him."

-    I John 2:3-5, NLT

cannot see inside your heart like God can. He knows when you have made the decision to ask Jesus to forgive you and be with you always. But mom and dad can't see that. Your grandmother and grandfather can't see that. So we join a church to show that we are part of God's family. We get baptized to show them that we believe in Jesus. We give them a sign, so they can see the decision we have made in our heart.

Your outward expression is very important for two reasons. First, when you get baptized, join a church, or do anything that shows others you are a Christian, you are *setting yourself apart*. This may change how you are viewed. Your friends may begin to ask you questions about what you believe, what you can or can't do, or why you do things like pray or go to church. Your teachers, friends, and family will look at what you do and say closely...sometimes in a good way... sometimes in bad. That is why you must understand the second reason, and that is you are showing God you are *committing to live for Him*. When you express your commitment to God, you are showing obedience and love for Him. God rejoices in our obedience and rewards us with peace so that we can be confident and excited about expressing our love for Him to others.

Just like a marriage, you should be certain that you are ready for baptism or joining a church. Knowing when is the right time to do something can be hard. The best way to help you decide is to pray for God to show you if it is the right time. Just remember, it's not about how old you are, or what other people your age are doing. It's about the decision you're making in your heart. Are you ready to make a commitment to be in obedience with God? If your answer is "yes" then you're probably ready. If you have questions, talk to your parents, a teacher, or even your pastor.

Don't just do it because someone told you that you should. Church families are really cool, and baptism is a great thing to celebrate. But the best thing is your inward decision, making the commitment in your heart to live for Jesus and *do what God wants you to do*.

Just so you know, I did finally get another wedding ring. Yeah, it's just a sign. But it's still cool, and I still marvel at its beauty and what it stands for. And yes, sometimes I still just like to hold up my hand and show off my beautiful ring.

GOD SAVES US TO SERVE

I have always loved the parable of the talents. In this story, found in Matthew, the ruler goes away and leaves three of his servants with different talents. I like the use of the word "talent" because it reminds me of the abilities God has given to us. In this parable, a talent is a kind of money. So even though there are different kinds of talents, the word is a good choice. Money is something of value, and from the way God sees it, so are the talents, gifts, and abilities that He gives us.

In this story, the ruler returns to see what the three servants accomplished while he was away. The one that he had given 5 talents doubled his, and returned 10. The servant who had 2 talents also doubled his, and returned 4. However, the servant with one talent was so afraid of losing his talent, that he hid his in the ground. Therefore, when the master returned, he had not used his talent at all, and he had no increase to show for the trust that the master had placed in him.

Matthew 25:14-30

It is the same with us, when God calls us out and saves us. He is placing His trust in us. We always have the choice to be like the first two servants, to invest what God has given us, and multiply it. Or we can take our talents and hide them so that no one ever knows that we follow Christ, and that our lives are supposed to be about Him. But we are called out to use our talents to build His kingdom. One way we can *honor the trust God has* in us is *by serving others*.

How can you honor God's trust?

One child that I know did not hide his talent, but multiplied it greatly to help those in need. During Children's Worship at church, the children participated in service project that involved collecting money to help those in need in our own home state, as well as across the country, and around the world.

One project was in support of Malaria No More. Malaria kills more children in Africa than any other single disease. Literally, one child dies every 30 seconds from this preventable and treatable disease. Malaria is caused by a parasite that gets into the human body by the bite of an infected mosquito. Malaria No More buys nets to protect people from mosquito bites. Their goal is to provide a mosquito net for every man, woman, and child at risk of malaria in Africa.

Never too Young

"Teach these things and insist that everyone learn them. Don't let anyone think less of you because you are young. Be an example to all believers in what you say, in the way you live, in your love, your faith, and your purity. Until I get there, focus on reading the Scriptures to the church, encouraging the believers, and teaching them."

"Don't fail to use the gift the Holly Spirit gave you…"

-     I Timothy 4:11-14, NLT, NIV

Our children brought in their pennies, nickels, and dimes to help children who lived thousands of miles away because they understood the words of Jesus:

"I tell you the truth, when you did it to one of the least of these My brothers and sisters, you were doing it to Me!"

One child did more than just give at our church. He took the idea to school. While discussing a history lesson on yellow fever, he mentioned how his church had helped children dying of malaria. This discussion turned into a school-wide campaign. Hundreds of children were given an opportunity to help thousands of children in another country because one child understood he could make a difference.

We are saved to serve, to help those who cannot help themselves. We are to use our talents to help others. Whether that talent is money, or that talent is the gifts and abilities that God has given you, even as a child, *it only takes one person to change the world*.

GOD GAVE US COMMUNION TO REMEMBER...

I watch children of all ages take Communion. At times, there is one in the group that catches my eye. One day as we were taking Communion in Children's Worship, I noticed one young lady in the back. She had basically turned Communion into her own private tea party. She would take the wafer and daintily dip it into her little cup of grape juice, and then she would nibble the edge. Then she would dip it again and nibble again. In those few seconds that I watched her, it was as if she had tuned the whole world out. She didn't seem to be aware of anything going on around her as she fancily enjoyed her wafer and juice.

However, she didn't stop at that. As it came time to drink, she threw her head back and turned the juice cup totally upside down. Although it was obvious to me that she had gotten every single drop of grape juice from the cup, it was apparent that she was not so convinced. She stuck out her tongue and began to lick all around the insides of the cup. I had to immediately tap her on the shoulder and indicate my disapproval. If I hadn't been so stunned, I may have considered that she had gotten an extra special Communion cup. Maybe her cup contained something so spectacular that she had to have more.

Well, I don't know how spectacular the grape juice in her cup was that day, but I do know how spectacular the story is behind why we drink the grape juice and eat the wafer.

There is one reason and one reason alone that *we take Communion* or The Lord's Supper, and it is *to remember Christ*. 1 Corinthians 11 tells us "Do this to remember me." But 1 Corinthians also reminds us, "A person should take a careful look at himself before he eats the bread and drinks from the cup."

So I tell my children this simple process. While waiting for everyone to be served Communion, take the time to pray. Think about the things that you have done that week that didn't please God. Be honest with God about the things you have done, and ask Him to forgive you. This prepares your heart.

Then I remind them to remember Christ, as the Bible teaches us to. What are we remembering? We remember how Christ died on the cross for us.

Communion is a time to remember.

Jesus died in our place. He didn't do anything wrong. He never sinned. He took all the punishment that should have been for us. The Bible teaches us that on the night Jesus was arrested He prayed so hard that He sweated drops of blood. He was really struggling to do the right thing, to make the right decision. He knew what God sent Him to Earth to do, but He was human like us. He was being honest when He talked to God in the Garden, but He concluded that it was not what He wanted that was important. It was what God wanted.

Later, the Bible tells us that they beat Jesus. Have you ever seen the whip that Indiana Jones uses? Well imagine nine of those tied together. Then add pieces of bone, glass, and rock to each leather strap. Now, think of how much it would hurt to have someone hit you with that whip over and over. Then, think about them not being happy just to beat you with the whip, but to then beat you where you are bleeding.

Have you ever touched the stem of a rose, like maybe one that you gave your mom for Mother's Day, and it nicked your finger and started to bleed? Imagine what would happen if someone took a whole bunch of rose stems with thorns and made them into a circle. Imagine the pain you would feel when they pressed that thorny-stem circle down onto your head. Then imagine that while all of this is happening to you, you are reminded that you are being punished for stuff that other people did, not things you did. You are innocent. You haven't done anything wrong.

Now, imagine that the same people drag you from place to place where other people say that they saw you do things that you didn't do, or they heard you say things that you didn't say. In addition, your friends that you hung with all the time, say that they never knew you, and they all run and hide so that they don't get in trouble too. And they leave you to go through all of this by yourself. Lastly, imagine that you know you are going to be nailed to a cross to die, and they make you carry that cross yourself. On top of that, they curse at you, spit on you, and pull out your hair. But you go through all of this torture, pain, and even death, because you love someone so much that you'd rather go through all of this in their place, so that they will never have to go through the same pain, and can have a chance to live a life filled with peace, love, and joy.

remember...

That is what we remember when we take Communion. Jesus didn't have to imagine all of these things because they happened to him. He died a horrible death because He loved us so very much that He wanted us to have a chance to live a life filled with peace, love, and joy. And to have a chance to live forever with God in Heaven, Jesus took our place. So Communion is a time to thank Him for what He has done for us, remembering what a great thing He did, and how much He loves us.

Next time you're given Communion...*remember*.

GOD GIVES YOU A STORY TO TELL

About 10 years ago, I was pregnant with twins. They were born too early, before they had enough time to grow and develop. I was so sad when they died, but I also believed that they were in heaven with God. I knew that there were other people who had gone through difficult things in their lives too...maybe their parents got a divorce, someone in their family was very sick, or maybe even their parents or grandparents had died.

When hard things happen to us, sometimes we wonder, "Why didn't God do something?" It was hard for me at first to understand how God could love me and why this bad thing had happened to me. But in time, I stopped being mad at God and saw how He was with me all that time, even when I was very sad. He was still with me.

After that, God gave me three more children. They didn't take the place of the two that had died, but I saw how God was still looking out for me. God still loved me.

I started writing down all the things that God had shown me when I was so sad. I took the time to think about all the good things God had done in my life. God put lots of people in my life that prayed with me and for me, and He helped me not be mad anymore.

I knew that there were lots of other people that had bad

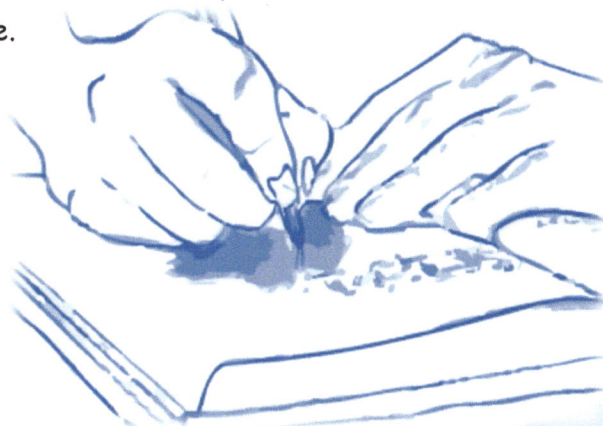

things happen to them, and maybe they didn't have people to pray with them and for them. Maybe they were so sad that they couldn't see how much God loved them and was still doing good things in their lives all around them.

Praise and Proclaim

"Oh, give thanks to the LORD! Call upon His name; Make known His deeds among the peoples! Sing to Him, sing psalms to Him; Talk of all His wondrous works!

"Sing to the LORD, all the earth; Proclaim the good news of His salvation from day to day. Declare His glory among the nations, His wonders among all peoples. For the LORD is great and greatly to be praised."

- 1 Chronicles 16: 8-9, 23-25, NKJV

So I took all of that stuff that I had learned and I put it in a book so that I could share it with other people to help them not be mad with God anymore. I wanted to help them see how much God loved them and all of the good things He was doing in their lives too.

One day, I was at a church speaking to people about all the things God taught me. Afterward a lady came up to me and slipped a piece of torn notebook paper into my hand and left in a hurry. She didn't even stay to talk to me. People were standing in line waiting to talk with me, so I couldn't run after the woman. But I instantly began to wonder what was on the notebook paper. I thought that I might have said something while speaking that made her cry or maybe made her mad, and she just couldn't face me.

When everyone had gone home, and I had some time to myself, I pulled out her note and began to read. She wrote about her daughter who had died. That day was her daughter's birthday, and she was having a very hard day. She wrote that she didn't want to talk to me because she thought she would start crying. She just wanted to thank me for taking the time to share my story. It had helped her get through the day.

I was glad that I decided to share what God had done in my life. Even though sometimes it was sad too for me to think about what happened to me, I was glad that I could share the things I had learned to help other people get to the point where they could be happy again.

God does that. No matter what happens to you, whether it is something good or something bad, *if you are willing* to tell other people, *God can use it* to make someone's life better. He gives us all a story, something that we share about our lives that will make a positive difference in the life of someone else.

Tell your story!

You never know what sharing your story might mean to someone else. If you are mad with God, or sad about something that has happened in your life, think about it this way. God loves you so and He knows that you love Him. When some not-so-good things happen in your life, God knows that you will be able to make it through. He also knows that there are some people that don't know how much He loves them. And when these same things happen in their lives, God knows that it might just break them. So He gives you an experience that you could share with others and help them make it through. By sharing your story with them, you let them know how much you love God, how much He

loves you, and how He helped you through your very hard time. By sharing your story, you show them how God can help them too.

No matter how old you are, or what has happened to you, you can tell your story. More often you will never know the positive difference you have made in the life of someone else. So, what are you waiting for? *GO TELL YOUR STORY!*

SALVATION "A·B·CS"

If you have already asked God to forgive you, and to give you that "Do-Over," that new life, then talk to Him, and see if there are areas of your life that you can still surrender to Him. Or discover what new things that He wants to teach you about who He is. Spend time with Him. Be real with Him about your life, and let Him lead the way. I love the word pictures from Psalm 36:7, "How precious is Your unfailing love, O God! All humanity finds shelter in the shadow of Your wings." You cannot be in someone's shadow unless you come very close to that person. Get closer to God, surround yourself in His love. Rest and relax in the shadow of His wings. Get right up close to Him, and let Him live His life in you and through you.

If you have never accepted His free gift of salvation, please consider it right now, this minute. This is the time. This is the place. He is waiting for YOU. It is truly the most important decision you will ever make. It will absolutely change the course of your life. If you don't know where to start, it is as simple as "A-B-C."

"A" stands for Admit. Take a look back over your life. There are probably some decisions that you have made that you thought were pretty good or at least not half bad. Things you have done that were generally good for others. However, you probably have those other kinds of decisions that were very bad, decisions that you regret. Everyone will find at least one thing in his or her entire life that they have done wrong. The Bible calls this sin. If you agree with God that even that *one* thing was wrong, that is the first step. By the way, the Bible says we all have sinned, so nobody gets a pass here. The issue is whether or not you are ready to "admit" your sin.

"B" stands for Believe. You have to move beyond just believing *about* God. Believing about God is believing that He made everything around us, that He is out there somewhere, that He knows all about you and wants you to do the right things.

You have to move beyond that and believe *in* God. Believing in God is believing that He is the only one who can save you from the punishment of sin, which is spiritual death or not having a relationship with God. It's believing that only He can give you new life. And it's believing that He loved you so much that He sent a perfect substitute to take your place. The substitute is Jesus. John 3:16 reminds us, "God loved the world so much that He gave His one and only Son. Anyone who believes in Him will not die, but will have eternal life."

You see, God put you here on this Earth, at this time and place for a purpose. Ultimately, He wants you to live with Him forever, but sin is what prevents that, and dooms all of us. There is no way for any man to rescue himself from this punishment. So God sent His Son, Jesus, who was God in the flesh, born of a virgin, the only human to live a perfect life without sin, to die on a cross and take the punishment of sin in your place. Then He brought His Son back from death so that everyone of us could have new life through Jesus. All you have to do is believe that God loved you enough to do all of this, and that's Step 2.

"C" stands for Confess. which simply means to tell, to make known, or to acknowledge. Romans 10:9-10 says, "If you confess with your mouth that Jesus is Lord and believe in your heart that God raised Him from the dead, you will be saved. For it is by believing in your heart that you are made right with God, and it is by confessing with your mouth that you are saved."

Confession is Step 3. Sometimes, writing your feelings down can help you sort things out. If you think you are ready to start your very own personal relationship with God through His Son Jesus, take some time to write a letter to Jesus and tell Him how you feel.

Write a letter to Jesus

As you write your "Jesus Letter," it will help you determine if you are ready to make a decision to follow Jesus. As you grow older, this letter will be a powerful reminder of what God was doing in your heart when you began a relationship with Him. It will also help you to share your faith story with others.

Find a quiet place to sit down, and get started. You can grab a pen and paper or use your book. If you need help, dictate your letter to a trusted adult who can write it down for you.

1. Write the date at the top of the page.

2. Begin with "Dear Jesus, You are . . ." ~ Who is Jesus to you?

3. Then admit, "I am a sinner, and I'm sorry for the times I . . ." ~ What have you done wrong?

4. Continue with what you believe, "Jesus, I believe You are . . ." ~ What do you believe about God?

5. Continue with, "I want to follow You because . . ." Tell Jesus why you are making this decision to follow Him.

6. Finish with your confession, "I confess this day and believe in my heart, that Jesus Christ is Lord."

7. Sign your name.

Put this letter in a safe place. You can use it to encourage yourself in your faith.

Today is _____ 20___

Dear Jesus, You are_____

_____

_____

I am a sinner, and I'm sorry for the times I _____

_____

_____

Jesus, I believe You are_____

_____

_____

I want to follow You because_____

_____

_____

"I confess this day and believe in my heart that Jesus Christ is Lord."

Sincerely,

_____ (sign your name)

Now that you have your letter, take some time, and simply talk with God. Let Him know that you believe, ask Him to forgive you, and ask Jesus to come live in your heart. That's it! That's Step 3. Remember, God saves in an instant. If you believe something in your heart, you shouldn't keep it a secret. So, go tell somebody! Go share your faith story!

GROW in HIM

Once you ask God to forgive you and save you, you become part of His family. You are a child of God. That's what it means to be born again. You get a new life, just like a baby. In the same way that we want and expect a newborn to grow into a toddler, then a little kid, a bigger kid, a teenager, and finally an adult, God wants us to grow spiritually as well. He wants you to continue to learn more about Him, and grow to be more like Him. There are four things you can do to grow. They are as follows:

1. **PRAY** (Talk to God every day, whether you are on your knees, riding your bike, sitting on the bus, or lying in bed. Talk to God, as often as you can.)

2. **READ and OBEY THE BIBLE** (Just like you wrote your letter to Jesus, God has written a letter to you, and it's called the Bible. Read the Bible, and you will learn what God has to say about lots of things. You'll also learn more about who God is. But it's not enough just to read it. Trust in the Holy Spirit to help you do what the Bible says.)

3. **TELL OTHERS ABOUT JESUS** (Share your faith story. Don't hide your talent in the ground. Let it multiply!)

4. **GO TO A BIBLE-BELIEVING CHURCH** (You can ask questions about things you don't understand, you can meet people who can encourage you in your faith, and you can learn more about how to please God.)

I'll leave you with one last verse, 1 John 1:9, "If we confess our sins, He is faithful and just to forgive us our sins and to cleanse us from all unrighteousness." That means once you become a child of God, when you sin, tell God about it. He already knows, and His grace is more than enough to give you another chance. Don't hide from Him, or think He will disown you. Be honest with Him. He loves you with an everlasting love....a love that will never end. That's why it is important to talk to God every day so you can develop a close relationship with Him. He promises in 1 John 1:9 that He will forgive you. Ask God to help you live a life that pleases Him!

If you have followed the A-B-Cs and given your life to Christ, I would love to hear from you, to celebrate with you and your family, and to pray for you. Consider sharing your story by e-mail at tina@tinavbryson.com, on the web at www.tinavbryson.com/10things.htm, or on Facebook at www.facebook.com/10thingsaboutGod.

CONCLUSION

Dear Reader:

I hope that this book has been a blessing to you. I hope it has helped to explain some of the things that parents, teachers, and friends have tried to teach you-- things you may have heard at church, or some things you just didn't understand. I hope that it has laid a foundation to help you understand a little more about the very special relationship God wants to have with you. I hope that it puts you way ahead of where I was when I was your age and trying to figure out what all of this God stuff really meant. But most of all, I hope you now see God in a new way. I hope that you now see that He pursues you, that He loves you so much, and He has sacrificed so much to have a close, personal relationship with you.

For Christ always,

Tina

About the Author

Tina V. Bryson is an author, speaker, and teacher with a passion for children's ministry. Her heart is to see children of all ages understand God's plan for them so that they live transformed lives for Him. She is also the author of *"Perfect in Weakness,"* a spiritual journey of God's faithfulness in our darkest hour. Tina and her husband Sebastian have four children.

www.ingramcontent.com/pod-product-compliance
Lightning Source LLC
Chambersburg PA
CBHW040230070426

42448CB00033B/56